SCIENCE BRAINERY

YOUR GO-TO INFORMATIONAL FRIEND.

ASHIRVAD A. NAIR

Made with ♥ on the Notion Press Platform
www.notionpress.com

Contents

BIOLOGICAL ADVENTURE

You're going to only learn the basic introduction to plants here. So, let's start.

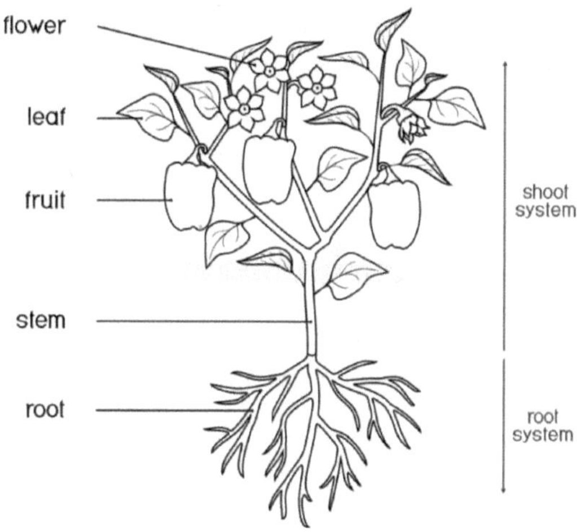

The parts of plants

The part above the soil is called **shoot system** and the part below the soil is called **root system.** The shoot system consists of flowers, leaves, fruits and the stem, whereas the root system, from its name consists of only the root.

THE TWO TYPES OF ROOT SYSTEMS

So, there are two types of root systems. The first one is called the **tap root system** and the second one is called the **fibrous root system**. In the tap root system, a single root called primary root comes out from the seed after germination. Later, smaller roots called lateral roots branch out from the primary root. But, in the fibrous root system, various roots branch out from the stem and they all look identical and have a bushy appearance.

Examples of tap root systems :- Beetroot, carrot, radish, turnip etc.

Examples of fibrous root systems :- Rice, wheat, maize, marigold etc.

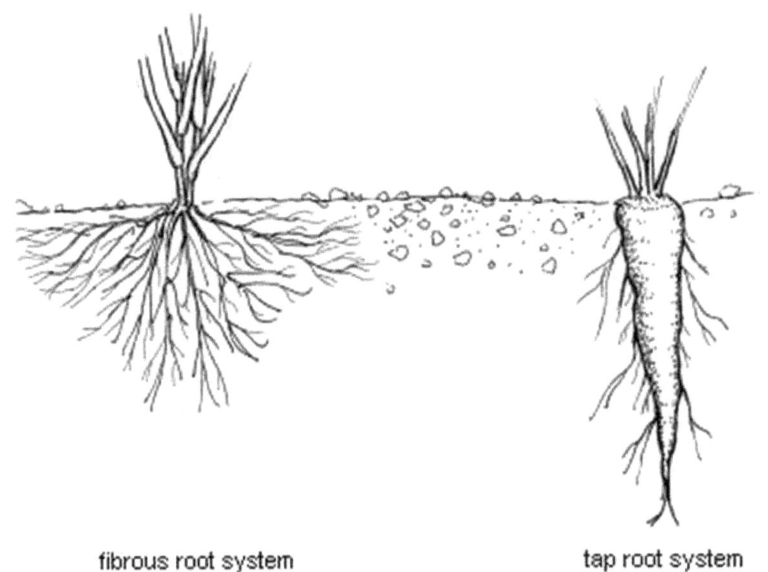

Fibrous root system and tap root system

PARTS OF A LEAF

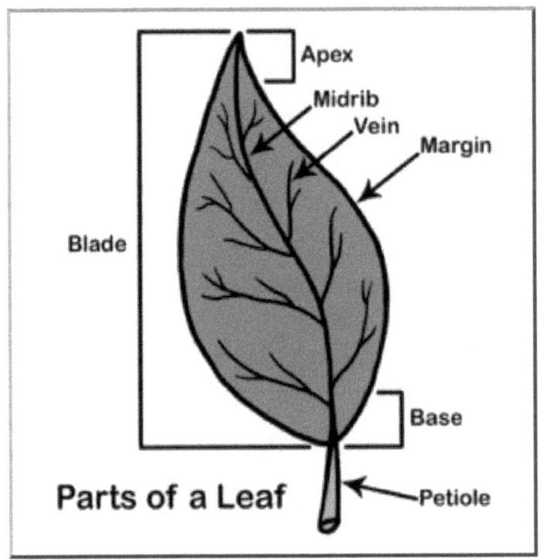

Parts of a leaf

Apex: tip of the leaf.

Margin: edge of the leaf.

Veins: carry food/water throughout leaf; act as a structure support.

Midrib: thick, large single vein along the midline of the leaf.

Base: bottom of the leaf.

Blade: the main light-collecting structure on a leaf; a large, broad, flat surface.

Petiole: the stalk that joins a leaf to the stem; leafstalk.

LEAF VENATIONS

The arrangement of veins in a leaf is called its **venation**. Like the root systems, there are two types of leaf venations; p**arallel venation** and **reticulate venation**. In parallel venation, the veins stand parallel to each other and the net-like design on both sides of the midrib is known as reticulate venation. Examples of parallel venation are banana, coconut, wheat, maize etc. Examples of reticulate venation are china rose, papaya, rose, mango, etc.

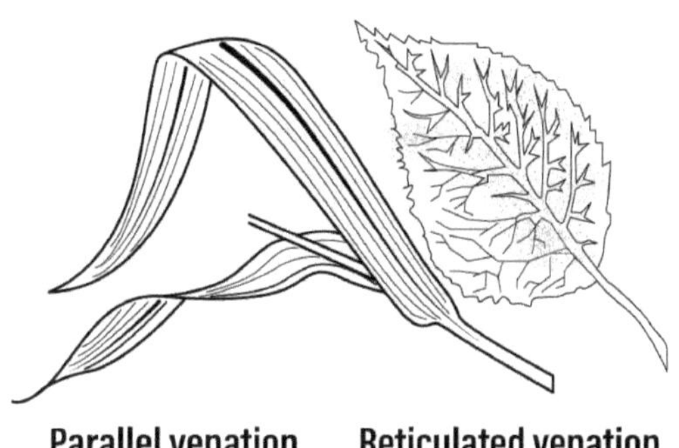

Parallel venation **Reticulated venation**

FLOWER

Now, we're going to look at the reproductive part of a plant, the **flower**.

Parts of a Flower

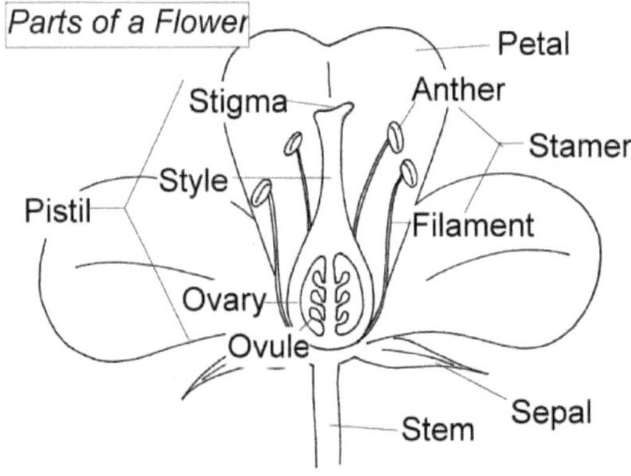

Parts of a flower

Stamen

Stamen is the male reproductive part of a flower. It has two parts. The knob-like structure called **anther** and thin stalk called **filament**. The anther produces a powdery substance called **pollen grains (pollen)**

Pistil

Pistil is the female reproductive part of a flower. It has three parts. Top part called **stigma**, enlarged base called **ovary** and the part that connects stigma and ovary, **style**. Inside the ovary, there are small ball-like structures called **ovules**. These later become seeds.

Petals

The colorful structures surrounding the inner part of the flower are called **petals.**

Sepals

The green leaf-like structures that protect the flower during its development are called **sepals.**

POLLINATION

Pollination is the transfer of pollen from an anther of a plant to the stigma of a plant, later enabling fertilisation and the production of seeds, most often carried by an animal or by wind. There are two types of pollination; self pollination and cross pollination.

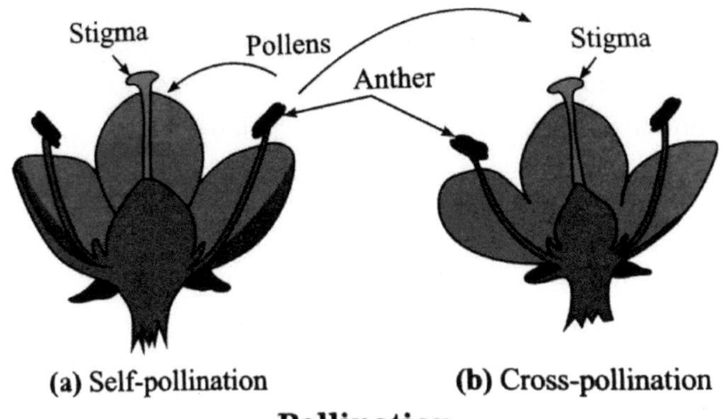

Pollination

Self Pollination

Self pollination can be of two types. (a) Same plant different flowers (b) Same flower.

Cross Pollination

The transfer of pollen from the anther of a flower from a plant to the stigma of a flow from another plant is called cross pollination.

TYPES OF PLANTS

There are many types of plants. They are :-

Herbs

Herbs are short plants with green, delicate and tender stems. They may have no or less branches. eg. tomato, wheat, grass etc.

Shrubs

Shrubs are plants with branches that grow near the stem. The stem of the shrub is tough but not particularly thick. Shrubs can be either **deciduous** or **evergreen**. eg. rose, lemon, cotton etc.

Creepers

Creepers are plants with weak stems that can't stand straight. They spread out on the ground. eg. watermelon, pumpkin etc.

Climbers

Climbers also have weak stems. They grow upwards by attaching itself to other plants or objects. eg. bottle gourd, bitter gourd etc.

Trees

Trees are plants with an extended stem, often known as a trunk, that support branches and leaves. Trees are essential for all living and non-living things on the planet. As they grow in size and diameter, they absorb carbon dioxide. eg. banyan, neem, peepal, oak etc.

FUNCTIONS OF ROOTS

- Roots absorb water and nutrients from the soil.
- They anchor the plant firmly.
- Roots bind the soil particles together preventing soil erosion.
- They transport water and minerals to the plant.

ROOT MODIFICATIONS

In some plants, roots are modified to perform special functions.

For storage of food

In some plants, roots are modified to store food in them.

For climbing

Some plants have climbing roots that help the plant to climb up on support walls or on trees. eg. money plant, beetal, black pepper etc.

For nutrition

Some plants have special roots called **parasitic roots** arising from their stem. They help the plant to absorb water and nutrients from the host plant.

For multiplication

Roots of some plants like dahlia and asparagus can develop into new plants.

For extra support

Some plants have extra roots that arise from their branches. These roots grow downwards and give extra support to the branches. eg. banyan tree.

FUNCTIONS OF STEM

- The stem bears flowers, buds, fruits and leaves.
- It holds leaves in position and ensures that leaves get enough sunlight for photosynthesis.
- The stem acts as pipelines and conduct water and minerals from the roots to the leaves. They also carry the food prepared by the leaves to other parts of the plants.
- In a few plants, green stems have chlorophyll and carry out photosynthesis.

STEM MODIFICATIONS

Stems of certain plants are also modified to perform special functions.

For storage of water

Stems of some plants such as cactus and jade swell up to store water in the stem.

For storage of food

Stems of plants like potato, ginger and onion are modifed to store food in their stem.

For manufacturing food

Stems of some plants become leaf like and flattened to perform photosynthesis.

For protection

Stems may be modified as thorns to protect the plants from being eaten by animals.

For support

Stems of some plants are modifed to form special structures called tendrils. These tendrils help plants like creepers which have weak stem to attach themselves to other. eg. passion fruit etc.

For multiplication

Rhizomes, bulbs and tubers also help in multiplication of the plants. Stem cuttings of plants such as jasmine, rose, sugarcane etc grow into new plants.

FUNCTIONS OF LEAVES

- Leaves perform photosynthesis.
- Extra food of plants are also stored in leaves.
- The exchange of gases take place through **stomata** seen on leaves.
- **Transpiration** takes place through the stomata.

LEAF MODIFICATIONS

For support

Leaves of some plants are also modified to form special structures called tendrils.

For protection

Leaves of some plants are modified to form spines for protection. Spines also reduce the transpiration rate.

WHY DO HUMANS DEPEND ON PLANTS?

Humans depend on plants in numerous ways. One reason we depend on plants is for consumption. Plants have the unique ability of producing their own food through a process called photosynthesis. In this process, plants are able to produce macromolecules such as carbohydrates that cannot be produced in animals or humans. In humans, the only to gain these macromolecules is to consume plant matter, or consume plant-eating animals (herbivores).

THE PLANT GROWTH STAGES

Plants' lives may be as short as a few weeks or months, but they go through distinct changes as they grow, just as people do. The stages that plants go through are from seed to sprout, then through vegetative, budding, flowering, and ripening stages. Similarly, the nutritional needs of people and plants change as they grow. This graphic shows how a plant develops (in this case, a tomato) and highlights the changing nutrient needs for plants as they grow.

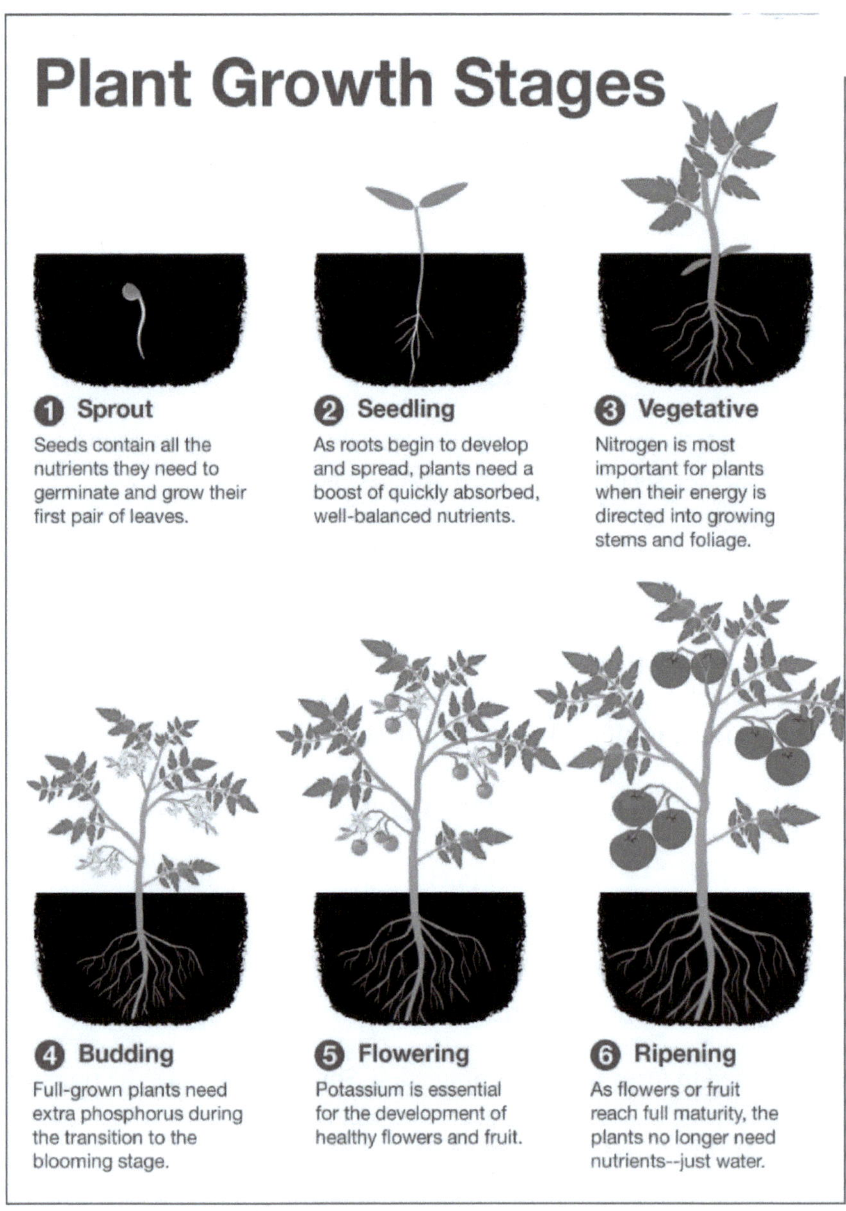

Now that you have read about the basics, here are some questions to solve

• • •

1. **What are the four functions of stem in plants?**
2. **What are the four functions of leaves in plants?**
3. **What are the two leaf modifications?**
4. **What is pollination?**
5. **What is cross pollination and self pollination? Explain.**
6. **Why do humans depend on plants?**
7. **What are atleast three root modifications?**
8. **What are the types of plants?**
9. **What is atleast one function of leaves in plants?**
10. **What are all the parts of a flower? Explain.**
11. **Explain the parts of a leaf.**
12. **Explain the two types of root systems.**
13. **What is a leaf venation? What are the two types of venations? Explain.**

• • •

Now we end off our biological adventure. But don't worry, there's more adventures for you to be in!

PHYSICAL ADVENTURE

What is physics? Physics is the branch of science that deals with the structure of matter and how the fundamental constituents of the universe interact. It studies objects ranging from the very small using quantum mechanics to the entire universe using general relativity. It is the scientific study of physical phenomena like the motion of matter and energy and force. It helps us to understand the world around us. Physics is the most fundamental part of science.

Here you'll only be learning the basics of physics.

WHAT IS MOTION?

- A brief explaination -

Motion is a change in position of an object over time. Motion is described in terms of displacement, distance, velocity, acceleration, time and speed. Without motion, life would cease to exist. You can look all around you and find something that has motion. For example, staring out of your window and the leaves on a tree are blowing back and forth. This is due to the effect of the wind. When we throw a baseball or drop a glass of milk and crashes to the ground. The physics of motion have pretty general rules that we all learned from elementary to make sense of the world. Motion has a lot that factors into the actually act. One is weight or mass between the objects. Weight and mass are two completely different terms and have different meanings. Weight is defined as the force of gravity on an object. So when we step on a scale to weigh ourselves, we are measuring the force of gravity against our body. Weight of an object equals the mass of an object time's acceleration of gravity. Mass of an object is the property that makes up the actual object. It's the matter we amass to create our body as a whole. So it's simply the space we take up in space. The difference between these two definitions is because they measure to completely different things.

So now we know what's the difference between mass and weight, we should see what actually sets objects in motion. One force that sets objects in motion would be gravity. We all experience gravity daily even when we wake up and sit up in bed. We have to go against gravity to lift up and sit upright. Gravity is considered one of the foundations of physics. It is always in action, when we walk, when we run, when we jump, etc. It's around us all day everyday. To help put gravity in perspective we have Newton's Laws. Newton's Law is the guidelines for objects in motion. .

Newton's first law states that an object will continue at rest or in motion in a straight line at a constant velocity unless acted upon by an external force. So if I dropped an apple from atop of a tree it should fall towards the earth at a constant speed with a direct path.

Now that you understood what motion is, let's MOVE on to the next topic.

GRAVITY

Gravity, also called gravitation, is a force that exists among all material objects in the universe. For any two objects or particles having nonzero mass, the force of gravity tends to attract them toward each other. Gravity operates on objects of all sizes, from subatomic particles to clusters of galaxies.

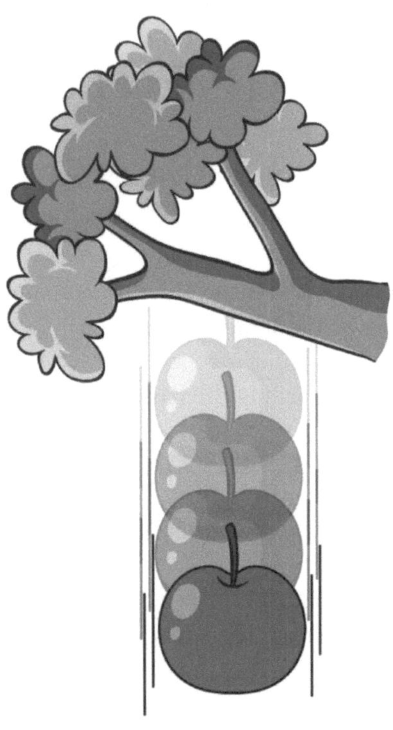

The apple falls because it is more attracted (gravitational force) to the Earth than it is to the tree.

BRIEFLY EXPLAINING GRAVITY

The intensity of force is directly proportional to the quantity of matter, so the things composed of more matter tend to be pulled with greater intensity of force. The mass of any object is the fundamental unit being used to measure the quantity of matter it that object. Therefore, the more heavy and matter in the object is, the higher gravitational pull will it put forth. Earth exerts a pulling force on the objects (Human beings) that move on its surface and as a result it pulls them back. As a matter of fact, the mass of earth is many times more than that of human beings so the pull from any walking individual is not significant enough to mobilize the Earth .On the other hand, the pulling impact of earth is strong enough to fall any walking individual flat. Gravity not only depends on the mass but it also largely depends on the distance, means how far an individual is from something. This is the reason why human beings are stuck to earth, rather being pulled away by Sun which is many times larger than the gravitational pull of the earth .

Gravity is considered as one of primary forces and helps in determining the weight of any object. The concept of gravity can be elaborated as, if an individual tries to examine his weight on a weighing machine than the scale reading indicates the amount of gravity being applied on his body. The universal formula for finding weight is: W=mg (w stands for weight, m for mass and g for gravity). The value of gravity is constant on the surface of Earth and it precisely taken as 9.8m/s2 .

Previously, scientists were of the view that massive objects tend to accelerate faster toward the ground. Recent researches and experiments have openly negated this assumption. The reason behind slow fall of a feather compared to a ball is due to the impact of resistance caused by air. The air resistance imposes its affect in direction opposite to that of gravitational pull. The renowned physicians, Isaac Newton put forwarded his famous Universal Law of Gravitation in 1680's, and he stated in this theory that gravity exerts it pull on matters and it depends on mass as well as distance. Every entity on this universe exerts a pull on other entity, and this pull is proportional to product of masses (of both the entities) and inversely proportional to the square distance between both the entities

. Fg = G (m1 · m2) / r2

- Fg stands for gravitational force
- m1 and m2 are the masses of both entities
- r stands for the distance between the entities
- G stands for universal gravitational constant

Newton's equation has been proved to be very instrumental in finding and predicting the behavior of the solar system. The Newton's work got published in the year 1687, and it was considered as the best possible elaboration and explanation. Later on, Einstein put forth the theory of relativity in the year 1915, he narrated that gravity is not merely a force; instead it is the outcome of the matter warps space –time. One of the assumption given by his theory is that light tend to bend around heavy object .

Technically, it is believed that gravity exists everywhere; it is gravity that provides the path to the orbits of the revolving planets of solar systems and also to galaxies. The source of gravity for the entire solar system is sun, it reaches everywhere and it is so powerful that it goes beyond the solar system. Gravity plays a pivotal role in ensuring the correct orbital shape and rotation of planets. The artificial satellite and the moon are under the influence of the gravity of earth and it keeps them in their designated orbits. It is an established fact that the influence or intensity of gravity decreases as the distance increases, so it is quite likely to have lesser influence of gravity at some far away point . The weightlessness experienced by the astronauts is not backed by the fact that they are at some far point. The fact is, astronauts tend to feel weightlessness is due to their location in contrast to that of spaceship. On the surface of earth humans feel weight because of the pulling gravitational effect that pulls them down, whereas ground prevents them from falling. Humans are being pressed against the ground so therefore don't feel weightlessness. Any spaceship orbiting around the earth is slowly and gradually falling to earth. Therefore, astronauts and the spaceship are falling at almost the same speed. The astronauts don't face any kind of pressing so they subject to weightlessness phenomena .

Among all other forces that naturally exist, gravity has been considered as the only universal one. Gravity affects and is being affected by all energy carrying objects and is in close contact with the shape of space time. The universal facet of gravity is clearly understood by the fundamental equation used for gravity, as it establishes very intimate resemblance with that of hydrodynamics and thermodynamics laws. A justified explanation of this fact has not been provided so far by the scientists. Gravity governs at very larger distances on the other hand it is weak at smaller readings. In reality, the fundamental laws of gravity have been experimented up to distance of millimeters. Scientists have revealed this fact that it is very difficult to combine gravity with the mechanism of quantum as compared to any other force that naturally exists. The struggle to combine gravity at microscopic scale with other natural forces is not practically possible. This approach may tend to causes many problems and will lead to uncountable puzzles and unsolved paradoxes. String theory has been successful in solving these to a certain limit, but not completely. It is still in process to learn about the solutions given by string theory .Many of the physicians strongly believed that the geometry of space time and gravity are at the emerging end. The developments related to string theory have shown numerous indications in solving all the problems. The most important clues come from ads/CFT and in more general term the open or close correspondence. This connection may lead to a certain level of duality between theories encompassing gravity and also those without gravity. Moreover, it proves that gravity can come out from as small as microscopic explanation that even have no idea about its presence .The universal nature of gravity strongly supports the idea that its presence needs be to understood with the help of general set of principle. These principles should be completely independent of any peculiar detail of the basic microscopic theory .

Despite large number of attempts in modern physics through theoretical models as well as practical experiments, the physicists are unable to give any final conclusion on the nature of gravity. There are seven major mysteries about gravity that need further research. The present models of physics including quantum mechanics and sting theories are not sufficient to describe the exact behavior of gravity. The major seven mysteries of gravity which are still unsolved are mentioned in subsequent paragraphs .

The first and the primary unresolved issue is regarding the nature of gravity. Gravity is considered among the four fundamental natures; however, it is undefined by any theoretical of practical model unlike other four fundamental forces. Gravity was initially defined by Newton which was considered final until Einstein changed it and its still not final. The second unresolved issue with the gravity is that the gravity has only one direction. The other forces and natural phenomena have two sides/directions. Like magnetic poles but gravity can only pull and never pushes. Scientists and physicists are unable to detect any object in the vastness of universe that could prove existence of gravitational push phenomena.

The third mystery about gravity is that the gravity is a very weak force when compared to other three fundamental forces in relation to the object size; however, this weak level of gravity can be easily detected in the vastness of universe. Gravity is undetected at small objects even at little distances but can be observed at larger distances for large objects. The fourth unresolved issue with gravity is that it has such a fine tuned value that even a slight change in the value of gravity would disrupt the systematic existence of the universe and all the objects within it including galaxies, stars, and planets.

The fifth mystery about gravity is that it is a basic necessity for existence of life. Although humans have gone to space and experienced zero gravity but it is not possible to live a healthy life without gravity. Humans and animals suffer from various medical problems including weakness of bones in absence of gravity. The sixth mystery of gravity is that despite high level of technologies, humans are still unable to make any antigravity material. Antigravity is a dream to be achieved by humans and there is still no clear indication that humans would be able to achieve it in future or not.

The seventh and final mystery of gravity is that, when gravity is explained through modern theories including relativity and quantum mechanics, a number of issues and paradoxes start to appear which remain unresolved. All these mysteries of gravity make the concept of gravity a much complex one as compared to what is presented through classical physics.

Gravity is experienced by everyone and is considered a simple force occurring in nature. However, it is not at all as simple as it seems. The classical concept of gravity as defined by Isaac Newton has been successful in the past but after various advancements including introduction of theory of relativity and quantum mechanics, gravity has become a serious topic of debate among physicists. Once gravity has been fully understood by physicists, it may become possible for us to make antigravity material and overcome or generate gravity wherever needed.

I guess that was a little more than basic level, was it? But, your physical adventure also is ending here. Let's move on to your next adventure!

PLANETARY ADVENTURE

From my opinion, this is my favourite topic to study about.

THE PLANETS

Sorry Pluto... But, Our solar system is made up of a star—the Sun—**eight planets**, yes, I said eight. 146 moons, a bunch of comets, asteroids and space rocks, ice, and several dwarf planets, such as Pluto. The eight planets are Mercury, Venus, Earth, Mars, Jupiter, Saturn, Uranus, and Neptune.

The eight planets in order

MERCURY

Mercury is the first planet that is the nearest to the sun. It is the fastest planet in the solar system, speeding along at about 29 miles per second and completing each orbit around the sun in just 88 Earth days. Mercury is also the smallest planet in the solar system, measuring just 3,032 miles wide at its equator.

VENUS

Venus is the second planet from the Sun and is Earth's closest planetary neighbor. It's one of the four inner, terrestrial (or rocky) planets, and it's often called Earth's twin because it's similar in size and density. These are not identical twins, however – there are radical differences between the two worlds.

EARTH

Earth, ofcourse is our home planet. It is a world unlike any other. The third planet from the sun, Earth is the only place in the known universe confirmed to host life. With a radius of 3,959 miles, Earth is the fifth largest planet in our solar system, and it's the only one known for sure to have liquid water on its surface.

MARS

Mars is the fourth planet from the sun. It is a terrestrial planet which means that like Earth it has a hard, rocky surface you can walk on. Mars is called the Red Planet because of all the iron in the soil. This iron rusts and as a result, makes the surface and atmosphere look red.

JUPITER

Jupiter is the largest planet in the solar system. Jupiter is so big that all the other planets in the solar system could fit inside it. More than 1,300 Earths would fit inside Jupiter. Jupiter is the fifth planet from the sun

SATURN

Saturn is the sixth planet from the Sun and the second-largest planet in our solar system. Like fellow gas giant Jupiter, Saturn is a massive ball made mostly of hydrogen and helium. Saturn is not the only planet to have rings, but none are as spectacular or as complex as Saturn's. Saturn also has dozens of moons.

URANUS

Uranus is the seventh planet from the Sun, and has the third-largest diameter in our solar system. It was the first planet found with the aid of a telescope, Uranus was discovered in 1781 by astronomer William Herschel, although he originally thought it was either a comet or a star.

NEPTUNE

Neptune is the eighth planet from the sun, the fourth largest, and a gas planet. It is named after the Roman god of the sea. Neptune is four times the size of Earth, and its day lasts a little more than 16 hours. Its year is about 165 Earth years.

THE ASTEROID BELT

The asteroid belt is a region of space between the orbits of Mars and Jupiter where most of the asteroids in our Solar System are found orbiting the Sun. The asteroid belt probably contains millions of asteroids.

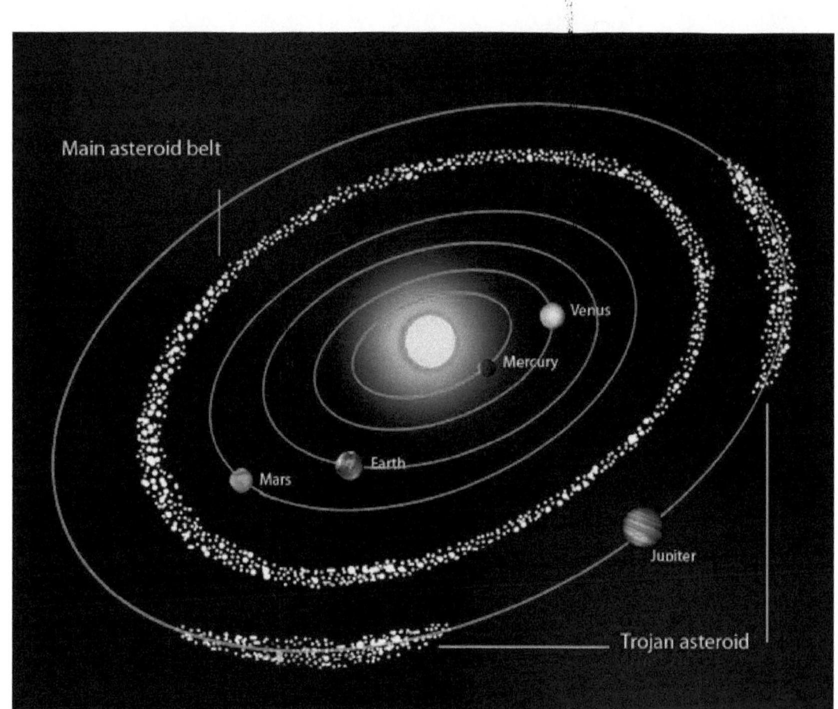

The asteroid belt

THE KUIPER BELT

The Kuiper belt is a circumstellar disc in the outer Solar System, extending from the orbit of Neptune at 30 astronomical units to approximately 50 AU from the Sun. It is similar to the asteroid belt, but is far larger—20 times as wide and 20–200 times as massive.

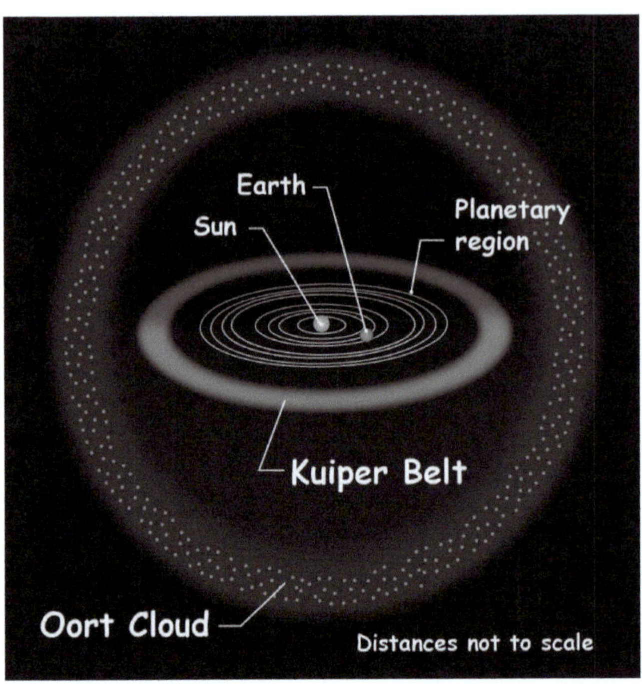

The Kuiper belt

ALIENS. REAL OR NOT?

Do aliens exist? It's a question that has spawned countless films and TV series, books, podcasts, artwork and conspiracy theories, but the question of whether or not we are truly alone in the universe remains unanswered.

Video transcript

We humans are obsessed with the idea that we're not alone, that somewhere out there in the universe, there are other life forms intelligent enough to communicate with us. Or maybe even destroy us. But if there is life out there, why haven't we heard from them yet? Or have we?

In 1961, the astronomer Frank Drake devised a calculation that predicted the existence of millions of civilisations out there among the stars. His Drake Equation used some fairly speculative assumptions, but it ignited our curiosity. Are we just one small part of a vast, cosmic zoo?

For physicist Enrico Fermi, the contradiction between this prediction and reality made no sense, and it came to be known as the Fermi Paradox. "Where is everybody?", he wanted to know. There could be an infinite number of answers to this question. Here are a few:

There is no paradox and we really are alone in the universe

Depressing though it may be, sometimes the simplest answers are the most plausible.

They're ghosting us

Maybe aliens have a policy not to interfere with less advanced cultures. A bit like the "prime directive", on Star Trek.

They don't live long enough to get in touch

This idea, known as the Great Filter, says that no advanced civilisation survives long enough to still be around when its neighbours are thriving. For us, threats like nuclear war, climate change or pandemics might spell our doom.

They're already here

From Roswell to The Lubbock Lights, there have been plenty of UFO sightings on Earth. Is the government hiding the evidence? Or is it fake news?

If there are any aliens watching, why not say hi in the comments below? Better yet, click like and subscribe. You might learn a lot about human behaviour from our videos. And, both Aliens and earthlings can get 20 per cent off a New Scientist subscription by using the code SAM20.

The silence from other civilisations is not for want of looking on our part. In 1960, Drake pointed a radio telescope towards two nearby stars, and waited for that hotline bling. Instead of intelligent messages he got a load of static, and interference from a secret military experiment. Nevertheless, the Search for Extra Terrestrial Intelligence, or SETI, was born.

Then, on 15 August, 1977, another SETI program recorded a brief radio signal coming from the direction of Sagittarius. Jerry Ehman, the astronomer analysing the data at 'Big Ear' telescope that day, was so excited that he wrote "Wow!" in the margin of the print-out. So far, that 'Wow!' signal stands as our most hopeful sign of alien communication yet.

But listening for ET to phone home is a bit like standing by the payphone, waiting for the phone to ring. Maybe we have to make the first move.

Drake tried this too, sending a radio message towards globular star cluster M13. Known as the Arecibo message, it showed potential aliens our DNA structure, solar system and some of the biochemicals of earthly life. We also tried sending aliens a greeting onboard space probes Pioneer 10 and 11, which included images of humans and directions to our planet.

Then, in 1977, we sent a more ambitious message, full of audio and images, on board the Voyager probes. We even included a mixtape of some of our favourite tunes, including Bach, Beethoven, Louis Armstrong and Chuck Berry, to impress any friendly lifeforms – or at least the ones with ears. And if that didn't entice them to get in touch, maybe this illustration of humans eating, licking and drinking might?!

By now aliens might know quite a lot about us – including what we need to survive and how to reach us. And you might be wondering whether it's such a good idea to be broadcasting our location all over the galaxy, when we don't know who's listening.

"It's like shouting in a forest before you know if there are tigers, lions, and bears", says Dan Werthimer, a SETI researcher. Worse still, these bears might have planet-destroying interstellar missiles.

Putting aside the question of whether we should try to contact aliens, are we even doing it right? These alien societies might be millions of years old, so to them our radio messages might look as outdated as flip phones, pagers, and, well, radio.

Plenty of other ways to look for life have been tried or suggested. Here are a few options:

Look for biosignatures

Over time, the biochemistry of billions of creatures can transform a world in distinctive ways. By analysing the wavelengths of light that come through the planet's atmosphere, we can look for gases that suggest the presence of life.

Look for technosignatures

if we do see signs of life on another planet, that won't tell us whether it is inhabited by mindless green slime or sentient city-builders. A more reliable approach to look for kindred spirits is to look for chemicals that only intelligent civilisations can produce.

City lights

Aliens with night vision as limited as ours might festoon their built environment with artificial lighting. So maybe we could spot cities lit up on the planet's dark side.

Aliens on the go

Alien spaceships might be easier to spot than their home planets. Maybe, we can look for high-powered lasers used to push optical sails, or intense plumes of light generated by an antimatter engine, like those imagined in Star Trek.

Megastructures

Aliens might have built gargantuan engineering projects like Dyson spheres, named after the physicist and engineer Freeman Dyson, which are swarms of devices around a star that harvest its energy, and nothing to do with vacuum cleaners

The Ozymandias Effect

The ruins of advanced civilisations doomed by their own technology might have their own telltale signatures. If we find dead civilisations before we find living ones, that might not bode well for our own future.

If we do make contact with aliens, it would be a decisive blow to the idea that humans are the centre of the universe. But how much cooler would it be to discover that we're just one branch of a vast galactic tree of life?

ESCAPE VELOCITY OF THE EARTH

Escape velocity is defined as the speed at which an object travels to break free from either the planet's or moon's gravity and leave without any development of propulsion. At Earth's surface, if atmospheric resistance could be disregarded, escape velocity would be about 11.2 km (6.96 miles) per second.

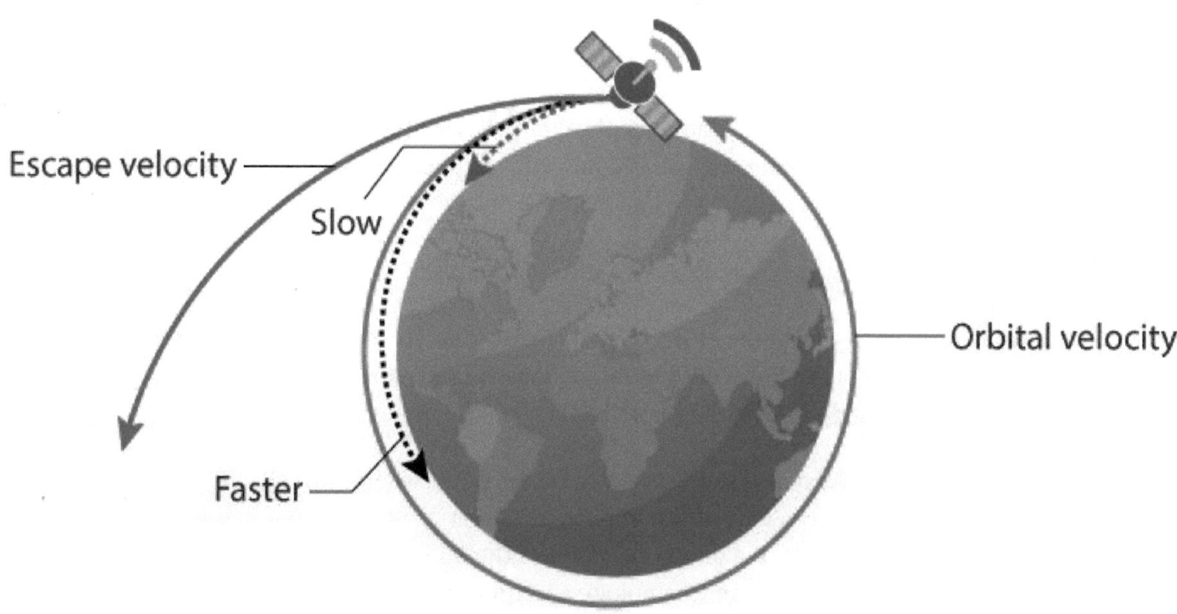

GALAXIES

A galaxy is a huge collection of gas, dust, and billions of stars and their solar systems. A galaxy is held together by gravity. Our galaxy, the Milky Way, also has a supermassive black hole in the middle.

ALL THE GALAXIES :-

All of the identified galaxies are Milky Way, Andromeda Galaxy, Sombrero Galaxy, Whirlpool Galaxy, Hoag's Object, Messier 82, Messier 83, Centaurus A, Pinwheel Galaxy, Large Magellanic Cloud, Tadpole Galaxy, Triangulum Galaxy, Cartwheel Galaxy, Sculptor Galaxy, Cosmos Redshift 7, Small Magellanic Cloud, Black Eye Galaxy, Comet Galaxy, NGC 4151, Messier 81, Messier 63, Circinus Galaxy, NGC 6946, Mayall's Object, Sculptor Dwarf Galaxy, Wolf Lundmark–Melotte, NGC 4565, Malin 1, GN-z11, NGC 4860, EGSY8p7, SSA22–HCM1, MCG+01-02-015, CEERS-93316, z8_GND_5296, 4C 41.17, SXDF-NB1006-2, NGC 7619, EGS-zs8-1, GN-108036, BDF-3299, Sagittarius Dwarf Spheroidal Galaxy, IC 10, NGC 4194, 3C 295, HCM-6A, Messier 99, Canis Major Overdensity, RD1, Baby Boom Galaxy and AM 0644-741.

But, a detailed theoretical simulation predicted far more faint, small galaxies than we've seen, upping the expected total to closer to 2 trillion. But recent observational evidence shows that even that estimate is far too low. Instead, there are between 6 and 20 trillion galaxies out there.

About our galaxy, The Milkyway

The idea that each star is a sun, many with their own solar systems, is a powerful reminder of the immense scale of the cosmos. However, the distances to stars in our galaxy are tiny in comparison to distances to other galaxies.

Since antiquity, observers have noted the existence of nebulous stars; diffuse smudgy or cloudy looking stars. Some of them turned out to be what we now know as nebulae, the places where stars form. Many turned out to be something else entirely. It wasn't until the 1920s when it was confirmed that many of these nebulous stars were in fact completely different galaxies, whole other sets of billions of stars like the Milky Way, far beyond our own.

We now know the Milky Way is but one of the billions of galaxies in the universe. Looking back at how astronomy developed this concept over time one can see how philosophers and scientists struggled with comprehending the nature of galaxies, and thus the enormity of our universe.

The Milky Way Resolves into More Stars

To the naked eye it is unclear exactly what the Milky Way is. In ancient Greece, the atomist philosopher Democritus had proposed that the bright band of light might consist of distant stars. The atomists' views were eclipsed by Aristotle's perspectives on the universe.

In Aristotelian Cosmology, the Milky Way was understood to be the point where the celestial spheres came into contact with the terrestrial spheres. One of the important observations Galileo noted in his 1610 Sidereus Nuncius was that, under the view of a telescope, parts of the Milky Way resolved into a cluster of many stars. Once again a weakness in Aristotelian Cosmology was found - the Milky Way wasn't the result of interactions between the terrestrial and celestial spheres. Galileo's observations demonstrated the Milky Way was a massive grouping of individual stars, planets and other nebulous elements.

Island Universes and External Creations

In 1750, English astronomer Thomas Wright, published An original theory or new hypothesis of the Universe. In this book, Wright speculated that the Milky Way was a flat layer of stars, a part of which which was our solar system.

Beyond this he suggested that many of the very faint nebulae "in all likelihood may be external creation, bordering upon the known one, too remote for even our telescopes to reach." The idea that the faint nebulae could be their own "external creations" suggested the universe was much large than previously imagined. In 1755, philosopher Immanuel Kant elaborated on Wright's ideas and referred to these faint nebulae as "island universes." Both the notions of external creations and island universes struggled to capture the implications of this new larger scale of the universe. Beyond the fact that our sun was a star, could nebulae be their own universes or completely separate creations?

Surveying the Milky Way

In the 1780s William Herschel surveyed the stars in a range of different directions. He found that the stars were much denser on one side of the sky than those of the other side.

His son John Herschel conducted a similar study of the sky in the southern hemisphere and found the same pattern. What they were seeing was the core of the Milky Way galaxy, where there is a much greater density of stars.

Herschel had placed our sun nearly at the center of the Milky Way; it wouldn't be until the 1920's when Harlow Shapley's demonstrated that our sun was far from the center of the Milky Way.

Andromeda and Other Nebulae

Nebulous stars have been observed for thousands of years. In 964 Islamic astronomer Al-Sufi had observed and recorded what he called "a small cloud" in an illustration of the constellation Andromeda. We now understand this description as the Andromeda galaxy. Only with the advent and refinement of the telescope was it possible to start to document different kinds of nebulous stars.

As already mentioned, Thomas Wright and Immanuel Kant had published their speculations that faint nebulous stars like this were in fact independent entities like the Milky Way. In the late 18th century Charles Messier compiled a catalog of the 109 brightest nebulae, which was followed by a William Herschel's much larger catalog of over 5,000. Even while documenting all of these nebulae it remained unclear as to exactly what they were.

Finding and Interpreting Red Shift

Studying the light spectrum of nebulae like Andromeda would ultimately provide the information about what exactly these objects were. A range of astronomers worked on this issue in the early 20th century. In 1912 astronomer Vesto Slipher studied the light spectra of some of the brightest nebulae. He was interested in determining if they were made of the kinds of chemicals one would expect to find in a planetary system.

Slipher found something very interesting - it is possible to calculate the relative speed and distance of a star or nebulae is moving by examining the light spectrum it gives off and seeing how much the indicators for elements have shifted into the blue or red color spectrum. Objects shifted blue are moving closer to us and red shifted objects are moving away from us. In Slipher's analysis, the spectrums for the nebula were shifted so far into the red that these nebulae must be moving away from the earth at speeds beyond the escape velocity of the Milky Way. Along with this evidence, in 1917 Herber Curtis observed a nova, the brightening of an exploding star, inside the Andromeda Nebula. Looking back over photographs of the Nebula he was able to document 11 more novae that were on average 10 times fainter than those of the Milky Way. The evidence was mounting to suggest that these nebulae were well outside the Milky Way.

In 1920, Harlow Shapley and Heber Curtis debated the nature of the Milky Way, nebulae and the scale of the universe. Using the 100 inch telescope at Mt. Wilson, Edwin Hubble was able to resolve the edges of some spiral nebulae to identify they were in fact collections of stars, some of which matched standard patterns that enable astronomers to calculate that the stars were too distant to be part of the Milky Way. Thus, the idea of the Milky Way as just one of many galaxies came to be the dominant scientific perspective.

Where the Earth was once understood to be the center of a relatively small universe we have come to understand it as one world orbiting one of the 300 billion stars in our galaxy which is itself just one of more than a hundred billion of galaxies in the observable universe. Even today it remains difficult to grasp just how tiny and small our planet is in the vastness of the observable universe.

Now, we have ended this adventure. Let's go on to the next one.

GEOGRAPHICAL ADVENTURE

Geography is the study of places and the relationships between people and their environments. Geographers explore both the physical properties of Earth's surface and the human societies spread across it.

WHAT EXACTLY IS GEOGRAPHY?

Geography as a university discipline got recognition in the early decades of the 19th century in the German universities and subsequently in the French and British universities.

During the period of evolution, geography, like all other sister social science disciplines, faced many philosophical and methodological problems. Geography did not develop as a well-regulated activity.

It followed a process of varying tensions in which tranquil periods, characterized by steady accretion of knowledge, are followed by crisis which can lead to upheaval within subject discipline and breaks in continuity. In each phase of tranquillity and crisis, geographical literature was and has been written with changing philosophies and methodologies; the philosophy and methodology being largely governed by the individual beliefs of the author, the political system, the social requirements of the people of the region and its economic institutions.

The last twenty-five years can be regarded as a period in which enormous geographical literature has been produced. This literature in the shape of books, research papers and monographs pertains to teaching, research, professional employment and pragmatic plans for the public and private bodies. Geography up to the Second World War, however, was regarded as a discipline providing general information about topography, relief features, weather, climate, mountains, rivers, routes, towns, cities and seaports.

Geography for most of the people was nothing but general knowledge. In the recent past, geographers have, however, adopted a new strategy in the restructuring of their courses and designed the syllabi around the theme of social welfare, making the subject the principal source of awareness of local surroundings, regional milieu, environmental pollution and world environment.

Geographers are venturing into the areas of environmental management and problems of pollution to make the social environment conducive for the proper development of individuals and societies. In order to achieve the welfare target, geographers are attacking social problems and exploring the causes of socio-economic backwardness, environmental pollution, and uneven levels of development in a given physical setting. Now, the main objective of geographical teaching and research is to train students in the analysis of phenomena, so that they can take up subsequently the problems of society as the fields of their research and investigation, thereby helping the local, state and national administration to overcome the regional and intra-regional problems.

The social problems are being tackled with approaches ranging from positive to normative, from radicalism to humanism, and from idealism to realism. In brief, geographers are increasingly concerning themselves with the problems of society, conditions of mankind, economic inequalities, social justice, and environmental pollution.

For the reduction of regional inequalities and for the improvement of the quality of life, the main concern of geographers is with what should be the spatial distribution of phenomena instead of with what it is. It is in this context that the spatial inequality in social amenities and living standards is investigated by geographers to trace the origin of disparity rather than to condemn injustice.

Historically, in the initial phases of its development, the main area of employment of geography students in the developed countries was teaching. In the Third World countries, geographers even today are not much actively involved in the process of planning and development. Regrettably, research had less important place in the geographical profession than in many of the social and physical sciences.

Moreover, the research done by individuals mainly remained confined to the libraries and has hardly been utilized for the purpose of planning. Unfortunately, the policy-makers in developing countries like India do not seem to be aware of the spatial dimensions of their problems of policies. Another reason is widespread ignorance of and even prejudice against geography particularly among the present generation of decision-makers whose opinions have been shaped by the experience of the previous generation school geography—when geography occupied a low place and was, as a subject, considered to be nothing more than general knowledge.

In fact, in most of the social fields, very little contribution had been made by geographers, and in the past they could not significantly suggest alternative strategies for the spatial organization of space. The last three decades have, however, seen some particularly important changes in the subject-matter, philosophy and methodology of geography. The major issues on which the geographers are concentrating include poverty, hunger, pollution, racial discrimination, social inequality or injustice, environmental pollution, and use and misuse of resources.

Some of the leading works which have been useful in the public policy making are: Geography of Crimes, Black-Ghetto, and Geography of Social Well-being. The quantitative revolution of the 1960s in geography gave to it some kind of intellectual vigour so essential for the rigorous analysis required in any public context and in the formulation of proposals for public policy.

It is an encouraging fact that now geographers all over the world are envisaging research on social problems with a welfare theme. They are working with a pragmatic approach to overcome the problems of inequalities. In fact, the objective of welfare geography is the evolution of the social desirability of alternative geographical state.

Scientific revolution entered in geography in the early 1970s. The pragmatists advocated the use of scientific methods (positivism) for finding solutions to human problems. It is with this intention that scholars like David M. Smith has adopted the welfare approach while discussing the problems and prospects of human geography.

The welfare geography has been defined differently by different scholars of geography. In the words of Mishan, "theoretical welfare geography is that branch of study which endeavours to formulate positions by which we may rank, on the scale of better or worse, alternative geographical situation open to society". While Nath expressed 'welfare geography' is that part of geography where we study the possible effects of various geographical policies on the welfare of society. In the spatial context, Smith defined welfare geography as the study of "who gets what, where and how".

The geographical 'state' or situation, in the sense used above, may refer to any aspect of the spatial arrangement of human existence. It may relate to the spatial allocation of resources, income, or any other source of human well-being. It may concern with the spatial incidence of poverty or any other social problem. The expression may also be used in desirable industrial location pattern, the distribution and concentration of population, the location of social service facilities,

transportation network, patterns of movement of people or goods and any other spatial arrangement which has a bearing on the quality of life as a geographically variable condition. And beneath them all, in the type of society—the economic, social, political structures that generate the pattern.

The welfare approach, nevertheless, has had different meanings in the different periods of human history. The humanist endeavours in various periods of different nations and societies like Jewish, Christians, Muslims, Confucians, Hellenistic, Scientific, Realists, Marxist and Existentialists, and many other forms of humanism appeared on the map of intellectual history.

The geographers who are mainly concerned with the problems of society and trying to formulate pragmatic proposals for public policy clarify the description and explanation of the phenomena. On the basis of such analysis they evaluate their plans and prescribe suitable strategies for balanced development.

Description involves the empirical identification of territorial levels of human well-being—the human condition. This is a major and immediate research area in which surprisingly little work has been done in India and in other developing countries. Explanation covers the how...It involves identifying the cause and effect links among the various activities undertaken in society, as they contribute to determining who gets what and where. This is where the analysis of the kind of economic, demographic and social patterns mentioned above logically fits into the welfare structure.

ALL THE CONTINENTS

ASIA

Asia is the world's largest and most diverse continent. It occupies the eastern four-fifths of the giant Eurasian landmass. Asia is more a geographic term than a homogeneous continent, and the use of the term to describe such a vast area always carries the potential of obscuring the enormous diversity among the regions it encompasses. Asia has both the highest and the lowest points on the surface of Earth, has the longest coastline of any continent, is subject overall to the world's widest climatic extremes, and, consequently, produces the most varied forms of vegetation and animal life on Earth. In addition, the peoples of Asia have established the broadest variety of human adaptation found on any of the continents.

The name Asia is ancient, and its origin has been variously explained. The Greeks used it to designate the lands situated to the east of their homeland. It is believed that the name may be derived from the Assyrian word asu, meaning "east." Another possible explanation is that it was originally a local name given to the plains of Ephesus, which ancient Greeks and Romans extended to refer first to Anatolia (contemporary Asia Minor, which is the western extreme of mainland Asia), and then to the known world east of the Mediterranean Sea. When Western explorers reached South and East Asia in early modern times, they extended that label to the whole of the immense landmass.

Asia is bounded by the Arctic Ocean to the north, the Pacific Ocean to the east, the Indian Ocean to the south, the Red Sea (as well as the inland seas of the Atlantic Ocean—the Mediterranean and the Black) to the southwest, and Europe to the west. Asia is separated from North America to the northeast by the Bering Strait and from Australia to the southeast by the seas and straits connecting the Indian and Pacific oceans. The Isthmus of Suez unites Asia with Africa, and it is generally agreed that the Suez Canal forms the border between them. Two narrow straits, the Bosporus and the Dardanelles, separate Anatolia from the Balkan Peninsula.

The land boundary between Asia and Europe is a historical and cultural construct that has been defined variously; only as a matter of agreement is it tied to a specific borderline. The most convenient geographic boundary—one that has been adopted by most geographers—is a line that runs south from the Arctic Ocean along the Ural Mountains and then turns southwest along the Emba River to the northern shore of the Caspian Sea; west of the Caspian, the boundary follows the Kuma-Manych Depression to the Sea of Azov and the Kerch Strait of the Black Sea. Thus, the isthmus between the Black and Caspian seas, which culminates in the Caucasus mountain range to the south, is part of Asia.

The total area of Asia, including Asian Russia (with the Caucasian isthmus) but excluding the island of New Guinea, amounts to some 17,226,200 square miles (44,614,000 square km), roughly one-third of the land surface of Earth. The islands—including Taiwan, those of Japan and Indonesia, Sakhalin and other islands of Asian Russia, Sri Lanka, Cyprus, and numerous smaller islands—together constitute 1,240,000 square miles (3,210,000 square km), about 7 percent of the total. (Although New Guinea is mentioned occasionally in this article, it generally is not considered a part of Asia.) The farthest terminal points of the Asian mainland are Cape Chelyuskin in north-central Siberia, Russia (77°43′ N), to the north; the tip of the Malay Peninsula, Cape Piai, or Bulus (1°16′ N), to the south; Cape Baba in Turkey (26°4′ E) to the west; and Cape Dezhnev (Dezhnyov), or East Cape (169°40′ W), in northeastern Siberia, overlooking the Bering Strait, to the east.

AFRICA

Africa is very rich in culture and has many diversions of that because within a country you can find dialects changing as well traditions from one country to another. After the rise of African nationalism, there is a cultural revival in the region. African people follows the number of distinct religions, they speak so many languages. Their living style is completely different from any other nation and western minority is heavily influenced by European culture.

Africa is the world's second-largest and second-most populous continent, after Asia in both cases. At about 30.3 million km2 (11.7 million square miles) including adjacent islands, it covers 6% of Earth's total surface area and 20% of its land area.[7] With 1.4 billion people[1][2] as of 2021, it accounts for about 18% of the world's human population. Africa's population is the youngest amongst all the continents;[8][9] the median age in 2012 was 19.7, when the worldwide median age was 30.4.[10] Despite a wide range of natural resources, Africa is the least wealthy continent per capita and second-least wealthy by total wealth, behind Oceania. Scholars have attributed this to different factors including geography, climate, tribalism,[11] colonialism, the Cold War,[17] neocolonialism, lack of democracy, and corruption.[11] Despite this low concentration of wealth, recent economic expansion and the large and young population make Africa an important economic market in the broader global context.

The continent is surrounded by the Mediterranean Sea to the north, the Isthmus of Suez and the Red Sea to the northeast, the Indian Ocean to the southeast and the Atlantic Ocean to the west. The continent includes Madagascar and various archipelagos. It contains 54 fully recognised sovereign states, eight territories and two de facto independent states with limited or no recognition. Algeria is Africa's largest country by area, and Nigeria is its largest by population. African nations cooperate through the establishment of the African Union, which is headquartered in Addis Ababa.

Africa straddles the equator and the prime meridian. It is the only continent to stretch from the northern temperate to the southern temperate zones.[18] The majority of the continent and its countries are in the Northern Hemisphere, with a substantial portion and number of countries in the Southern Hemisphere. Most of the continent lies in the tropics, except for a large part of Western Sahara, Algeria, Libya and Egypt, the northern tip of Mauritania, and the entire territories of Morocco, Ceuta, Melilla, and Tunisia which in turn are located above the tropic of Cancer, in the northern temperate zone. In the other extreme of the continent, southern Namibia, southern Botswana, great parts of South Africa, the entire territories of Lesotho and Eswatini and the southern tips of Mozambique and Madagascar are located below the tropic of Capricorn, in the southern temperate zone.

Africa is highly biodiverse; it is the continent with the largest number of megafauna species, as it was least affected by the extinction of the Pleistocene megafauna. However, Africa also is heavily affected by a wide range of environmental issues, including desertification, deforestation, water scarcity and pollution. These entrenched environmental concerns are expected to worsen as climate change impacts Africa. The UN Intergovernmental Panel on Climate Change has identified Africa as the continent most vulnerable to climate change.[19][20]

The history of Africa is long, complex, and has often been under-appreciated by the global historical community.[21] Africa, particularly Eastern Africa, is widely accepted as the place of origin of humans and the Hominidae clade (great apes). The earliest hominids and their ancestors have been dated to around 7 million years ago, including Sahelanthropus tchadensis, Australopithecus africanus, A. afarensis, Homo erectus, H. habilis and H. ergaster—the earliest Homo sapiens (modern human) remains, found in Ethiopia, South Africa, and Morocco, date to circa 233,000, 259,000, and 300,000 years ago respectively, and Homo sapiens is believed to have originated in Africa around 350,000–260,000 years ago.[28] Africa is also considered by anthropologists to be the most genetically diverse continent as a result of being the longest inhabited.[29][30][31]

Early human civilizations, such as Ancient Egypt and Carthage emerged in North Africa. Following a subsequent long and complex history of civilizations, migration and trade, Africa hosts a large diversity of ethnicities, cultures and languages. The last 400 years have witnessed an increasing European influence on the continent. Starting in the 16th century, this was driven by trade, including the Trans-Atlantic slave trade, which created large African diaspora populations in the Americas. From the late 19th century to the early 20th century, European nations colonized almost all of Africa, reaching a point when only Ethiopia and Liberia were independent polities.[32] Most present states in Africa emerged from a process of decolonisation following World War II.

AUSTRALIA

Australia is the only country in the world that covers an entire continent. It is one of the largest countries on Earth. Although it is rich in natural resources and has a lot of fertile land, more than one-third of Australia is desert.

Most Australian cities and farms are located in the southwest and southeast, where the climate is more comfortable. There are dense rain forests in the northeast. The famous outback (remote rural areas) contains the country's largest deserts, where there are scorching temperatures, little water, and almost no vegetation.

Running around the eastern and southeastern edge of Australia is the Great Dividing Range. This 2,300-mile (3,700-kilometer) stretch of mountain sends water down into Australia's most important rivers and the Great Artesian Basin, the largest groundwater source in the world.

EUROPE

Europe is the second-smallest continent. Only Oceania has less landmass. Europe extends from the island nation of Iceland in the west to the Ural Mountains of Russia in the east. Europe's northernmost point is the Svalbard archipelago of Norway, and it reaches as far south to the islands of Greece and Malta.

Europe is sometimes described as a peninsula of peninsulas. A peninsula is a piece of land surrounded by water on three sides. Europe is a peninsula of the Eurasian supercontinent and is bordered by the Arctic Ocean to the north, the Atlantic Ocean to the west, and the Mediterranean, Black, and Caspian Seas to the south.

Europe's main peninsulas are the Iberian, Italian, and Balkan, located in southern Europe, and the Scandinavian and Jutland, located in northern Europe. The link between these peninsulas has made Europe a dominant economic, social, and cultural force throughout the recorded history.

Europe's physical geography, environment, resources, and human geography can be considered separately.

Europe can be divided into four major physical regions, running from north to south: Western Uplands, North European Plain, Central Uplands, and Alpine Mountains.

NORTH AMERICA

North America, the third-largest continent, extends from the tiny Aleutian Islands in the northwest to the Isthmus of Panama in the south. The continent includes the enormous island of Greenland in the northeast and the small island countries and territories that dot the Caribbean Sea and western North Atlantic Ocean.

In the far north, the continent stretches halfway around the world, from Greenland to the Aleutians. But at Panama's narrowest part, the continent is just 50 kilometers (31 miles) across.

North America's physical geography, environment, resources, and human geography can be considered separately. Mexico and Central America's western coast are connected to the mountainous west, while its lowlands and coastal plains extend into the eastern region.

Within these regions are all the major types of biomes in the world. A biome is a community of animals and plants spreading over an extensive area with a relatively uniform climate. Some diverse biomes represented in North America include desert, grassland, tundra, and coral reefs.

SOUTH AMERICA

South America, the fourth-largest continent, extends from the Gulf of Darien in the northwest to the archipelago of Tierra del Fuego in the south. South America's physical geography, environment, resources and human geography can be considered separately.

South America's extreme geographic variation contributes to the continent's large number of biomes. A biome is a community of animals and plants that spread over an area with a relatively uniform climate.

Within a few hundred kilometers, South America's coastal plains' dry desert biome rises to the rugged alpine biome of the Andes mountains. One of the continent's river basins (the Amazon) is defined by dense, tropical rain forest, while the other (Parana) is made up of vast grasslands. With an unparalleled number of plant and animal species, South America's rich biodiversity is unique among the world's continents.

ANTARCTICA

Antarctica is Earth's southernmost continent, underlying the South Pole. It is situated in the Antarctic region of the southern hemisphere, almost entirely south of the Antarctic Circle, and is surrounded by the Southern Ocean. At 14.0 million km2, it is the fifth-largest continent in area after Asia, Africa, North America, and South America. About 98% of Antarctica is covered by ice, which averages at least 1.6 kilometres (1.0 mi) in thickness.

Antarctica, on average, is the coldest, driest, and windiest continent, and has the highest average elevation of all the continents. Antarctica is considered a desert, with annual precipitation of 8 inches along the coast and far less inland. There are no permanent human residents but anywhere from 1,000 to 5,000 people reside throughout the year at the research stations scattered across the continent. Only cold-adapted plants and animals survive there, including penguins, seals, nematodes, Tardigrades, mites, many types of algae and other microorganisms.

The coastline measures 17,968 km and is mostly characterized by ice formations.
Antarctica is divided in two by Mountains close to the neck between the Ross Sea and the Weddell Sea. The portion west of the Weddell Sea and east of the Ross Sea is called West Antarctica and the remainder East Antarctica, because they roughly correspond to the Western and Eastern Hemispheres relative to the Greenwich meridian.

About 98% of Antarctica is covered by the Antarctic ice sheet, a sheet of ice averaging at least 1.6 km (1.0 mi) thick. The continent has about 90% of the world's ice and about 70% of the world's fresh water.

Antarctica is white. it doesn’t get much sunlight, but what it gets is largely reflected back into space. this means that it keeps the world cooler by limiting the amount of light that is turned into heat. Antarctica is cold. In a way, it regulates the temperature of the rest of the planet by being a sink for heat.

This adventure also is going to be concluded here.

The Conclusion

I feel sad that all our adventures are ending here. I really hope you have learned a lot from this book.

Like that, we end off all the adventures we've been through..

The mission of this book really was to give out information. So, I really do hope that you understood.

9 798889 592846

Printed by Libri Plureos GmbH in Hamburg,
Germany